Abdelhafid Mimouni

Humans and Macaques: Regulation of Blood Pressure

Abdelhafid Mimouni

Humans and Macaques: Regulation of Blood Pressure

ScienciaScripts

Imprint

Cover image: www.ingimage.com

This book is a translation from the original published under ISBN 978-620-6-72740-8.

Publisher:
Sciencia Scripts
is a trademark of
Dodo Books Indian Ocean Ltd. and OmniScriptum S.R.L publishing group

120 High Road, East Finchley, London, N2 9ED, United Kingdom
Str. Armeneasca 28/1, office 1, Chisinau MD-2012, Republic of Moldova, Europe
Managing Directors: Ieva Konstantinova, Victoria Ursu
info@omniscriptum.com

Printed at: see last page
ISBN: 978-620-8-39668-8

Humans and Macaques: Regulation of Blood Pressure

Author: Dr. Abdelhafid Mimouni: Independent researcher in bioinorganic chemistry, holder of a doctorate in chemistry from the University of Paris XII (1997) and a Diplôme des Études Approfondies en Systèmes Bioinorganiques from the University of Paris XI (93), a Licence and a Maîtrise in chemistry (91, 92).

Summary: This book provides a comparative study of the mechanisms of blood pressure regulation in humans and macaques, highlighting their physiological and anatomical differences. It explores hormonal systems, such as the renin-angiotensin-aldosterone system, as well as the interaction between the sympathetic and parasympathetic nervous systems in the control of blood pressure. The impact of diet, stress and physical activity on cardiovascular health is also examined, highlighting the role of macaques as models for the study of human hypertension. Finally, implications for future research and translational research aimed at improving clinical interventions and prevention strategies in cardiovascular health are discussed.

Book outline :

Introduction

Blood pressure regulation is a vital process that maintains homeostasis within the body. A precise balance of blood pressure is essential to ensure adequate perfusion of tissues and organs, enabling them to function properly. Abnormalities in this regulatory system can lead to serious pathologies, in particular hypertension, which is a major risk factor for cardiovascular disease, stroke and kidney disease.

This subject is of particular importance not only for human health, but also for our understanding of physiological mechanisms in other species. Macaques, for example, are an interesting model because of their genetic proximity to humans and their physiological similarities. However, despite the progress made in research into blood pressure in humans, data on the regulatory systems in macaques remains relatively limited. This gap raises questions about adaptive variations and clinical implications for human health.

The aim of this comparative study is to highlight the similarities and differences in the blood pressure regulation systems of humans and macaques. By examining the underlying biological mechanisms, neural and hormonal responses, as well as the influence of environmental factors, this research aims to provide a fresh perspective on cardiac physiology and explore how this knowledge can enrich our understanding of cardiovascular disorders. Ultimately, this study aims to open up avenues for future research, both fundamental and applied, in bioinorganics and health.

Chapter 1: Cardiac anatomy and physiology

Comparison of cardiac anatomy between humans and macaques

Cardiac anatomy is a fundamental area of study for understanding the mechanisms by which blood pressure is regulated. The human heart is a hollow muscular organ divided into four chambers: two atria (right and left) and two ventricles (right and left). This structure ensures effective separation between oxygenated blood, which circulates in the systemic circulation, and deoxygenated blood, which circulates in the pulmonary circulation. The coordination of cardiac contractions is orchestrated by the cardiac conduction system, which includes the sinus node and the atrioventricular node.

Macaques, as non-human primates, have anatomical similarities with humans, but also important distinctions. Their hearts have a similar structure, with four chambers, but variations in the size and thickness of the myocardial walls can be observed. These differences are often linked to the environment in which they evolve and the physiological adaptations required to meet their metabolic needs. For example, the myocardial mass of macaques may be greater due to their frequent physical activity and social lifestyle, requiring increased blood pumping capacity (Dunbar, 2009).

In addition, coronary vessel morphology can vary between species, influencing myocardial perfusion and responses to exercise or stress.

Imaging and comparative anatomy studies show that macaques have more developed coronary arteries, allowing them to better support elevated heart rate and increased metabolic demand (Lindsey et al., 2014).

Heart rate and implications for circulation

Heart rate (HR) is a key indicator of cardiac performance and vascular health. In humans, the resting heart rate is generally between 60 and 100 beats per minute (bpm). In macaques, however, resting heart rate can vary considerably, often reaching 150 bpm or more depending on the species and physiological conditions (Hoffman et al., 2004). This significant difference has profound implications for blood circulation.

A higher heart rate in macaques makes it possible to increase cardiac output (the volume of blood pumped by the heart per minute), which is crucial for meeting their metabolic needs, especially during intense physical activity or in stressful situations. This ability to rapidly modulate HR is essential in environments where survival depends on responsiveness to predators or environmental changes (Harrison et al., 2013).

Heart rate and regulation by the autonomic nervous system

The high heart rate observed in macaques is the result of complex regulation orchestrated by the autonomic nervous system (ANS). This system, which functions involuntarily, is divided into two main branches: the sympathetic nervous system and the parasympathetic nervous system. Each of these branches plays a crucial role in modulating heart rate and cardiovascular responses.

Role of the sympathetic nervous system

The sympathetic nervous system (SNS) plays a central role in preparing the body to respond to stressful or emergency situations, often referred to as the 'fight or flight' response. This mechanism, of vital importance for survival, enables macaques (and other animals) to react quickly to potential threats, such as predators or social rivalry.

Activation of the sympathetic nervous system

When the macaque perceives a threat or engages in intense physical activity, sensory signals trigger activation of the SNS. This activation leads to the rapid release of neurotransmitters, notably noradrenaline, which is secreted by sympathetic nerve endings. Noradrenaline has a variety of effects on many of the body's systems, but its impact on the heart is particularly crucial.

Effects on the heart

1. **Increased heart rate**: Noradrenaline acts primarily on the sinus node, the heart's natural pacemaker. By increasing the rate of

depolarisation of the cells in the sinus node, noradrenaline increases the heart rate, enabling the heart to pump more blood per minute. This increase is essential to provide an increased supply of oxygen to the muscles and vital organs during periods of stress or activity.

2. **Stimulation of myocardial contractility**: As well as increasing heart rate, SNS also boosts the contraction force of the heart muscle, a phenomenon known as myocardial contractility. This means that each heartbeat pumps more blood, thereby increasing cardiac output (the total volume of blood pumped by the heart in one minute). This ability to adjust the force of contraction is vital in situations where maximum physical performance is required.

Importance in social interaction

As social animals, macaques live in groups where interactions can be frequent and sometimes conflictual. In this context, the ability to react quickly to social stimuli is crucial. For example, during confrontations with other members of the group, an effective sympathetic response enables a macaque to quickly mobilise the energy needed to defend itself or establish dominance. Studies have shown that this ability to rapidly adjust heart rate and force of contraction is particularly important in situations where decisions need to be made quickly (Reid et al., 2015).

Underlying physiological mechanisms

Activation of the SNS is not limited to the release of noradrenaline. It also involves hormonal mechanisms, such as the release of adrenaline by the adrenal glands. This hormone, which has similar effects to noradrenaline, further increases heart rate and myocardial contractility. Together, these hormonal and nervous actions prepare the body to cope with stressful situations.

In short, the sympathetic nervous system plays an indispensable role in preparing the macaque to react to threats and physiological demands. By increasing both heart rate and myocardial contractility, the SNS enables rapid adaptation to metabolic needs, ensuring survival and efficiency in complex social environments. These physiological responses are essential not only for individual survival, but also for the dynamic functioning of the social groups to which macaques belong.

Role of the parasympathetic nervous system

Although the sympathetic nervous system (SNS) plays a dominant role in activation responses, the parasympathetic nervous system (PNS) is equally essential for the management of post-stress physiological responses. Primarily mediated by the vagus nerve, the PNS is responsible for modulating bodily functions at rest and for recovery after periods of intense activity or stress.

Activation of the parasympathetic nervous system

When the threat or stress diminishes, the body activates the PNS, restoring a state of equilibrium. This activation is particularly important for managing the physiological effects induced by sympathetic activation. By increasing parasympathetic tone, the body initiates a recovery process that helps to gradually reduce heart rate and restore a state of calm.

Effects on the heart

1. **Reduced heart rate**: The parasympathetic tone has a direct effect on the sinus node, slowing down the rate of depolarisation. This leads to a reduction in heart rate, allowing the heart to return to a more normal rhythm after a period of stress. This heart rate reduction phase is crucial to avoid excessive fatigue of the heart muscle and to maintain healthy cardiovascular function.
2. **Improved recovery**: By promoting a return to a resting state, SNP helps to reduce stress levels on the heart. Adequate recovery is essential to avoid the harmful effects of prolonged sympathetic stimulation, such as hypertension and overload of the cardiovascular system. The regulation of heart rate by the PNS also helps to minimise the risk of long-term cardiovascular complications.

Protection against the deleterious effects of stress

Increased parasympathetic tone plays an important protective role. When the PNS is activated, it helps to counteract the negative effects of excessive sympathetic stimulation, such as increased blood pressure, inflammation and cardiac fatigue. This fine regulation is essential for maintaining homeostatic balance in the body, especially in stressful environments or during periods of prolonged stress.

Studies have shown that in macaques, adequate modulation of parasympathetic tone is associated with greater resilience to stress and improved cardiovascular health (Reid et al., 2015). Indeed, macaques able to better regulate their parasympathetic response show an increased ability to cope with stressful situations without compromising their cardiovascular health.

Interactions with the sympathetic nervous system

The PNS and SNS often work in opposition, but they must also work together to ensure a balanced physiological response. While the SNS activates the body to deal with emergency situations, the SNP helps to calm and regulate this excitement once the threat has passed. This interaction between the two systems is crucial for maintaining homeostasis and promoting optimal health.

In short, the parasympathetic nervous system plays a vital role in post-stress recovery and the modulation of cardiac functions. By lowering heart rate and facilitating recovery, the PNS helps to maintain physiological balance and protect the heart from the deleterious effects of prolonged stress. Understanding the importance of this system in the cardiovascular health of macaques provides valuable insights for research into stress regulation and general well-being.

Interaction between the two systems

The interaction between the sympathetic and parasympathetic nervous systems is essential for the effective regulation of heart rate.

1. **Stress response**: When the body is faced with a stressful situation, the sympathetic system is activated. This leads to an increase in heart rate and the force of contraction of the heart, enabling the body to react quickly to threats or physical demands.
2. **Recovery from stress**: Once the threat has passed, the parasympathetic system takes over. It helps bring the heart back to a resting state by lowering the heart rate and promoting recovery. This process is crucial in preventing overwork of the heart and maintaining optimal cardiovascular function.
3. **Dynamic balance**: This interaction between the sympathetic and parasympathetic systems creates a dynamic balance. This allows

the body to respond quickly to immediate needs while ensuring adequate recovery from periods of intense stress.

In short, the regulation of heart rate in macaques highlights the importance of harmonious collaboration between the two branches of the autonomic nervous system. This fine regulation is not only essential for physical performance, but is also crucial for the overall health of the heart and cardiovascular system. Poor interaction between these two systems can lead to health problems such as hypertension and other cardiovascular disorders.

Conclusion

The anatomical and physiological differences between human and macaque hearts, particularly in terms of heart rate, reveal mechanisms for regulating blood pressure and responses to environmental stimuli that are crucial to cardiovascular health.

1. **Implications of anatomical differences**: The cardiac structure of macaques, although similar to that of humans, shows significant variations that influence their cardiovascular function. For example, myocardial mass and coronary vessel morphology in macaques can be adapted to their specific metabolic needs, due to their active and social lifestyle. These anatomical distinctions may offer insights into how different species deal with physical and

emotional stress, and how these responses could be applied to understanding human cardiac pathologies.

2. **Heart rate and blood pressure regulation**: The higher heart rate in macaques, which can reach 150 bpm at rest, is a key element in the regulation of their blood pressure. Understanding how this higher heart rate is maintained and modulated by the autonomic nervous system may shed light on the mechanisms controlling blood pressure, a major risk factor for cardiovascular disease. This knowledge may also influence therapeutic approaches aimed at treating hypertension in humans.
3. **Responses to environmental stimuli**: The physiological responses of macaques to environmental stimuli, orchestrated by the interaction between the sympathetic and parasympathetic systems, underline the importance of balanced stress regulation. These mechanisms can be applied to better understand stress-related disorders in humans, such as anxiety or depression, which have direct consequences for cardiovascular health.
4. **Health applications and comparative medicine**: Comparative studies of the human and macaque cardiovascular systems can provide valuable insights for cardiovascular health research. For example, macaques are often used as animal models to test medical interventions and evaluate the efficacy of new treatments. By better understanding their physiological mechanisms,

researchers can develop more effective strategies for preventing and treating heart disease in humans.

In short, exploring the differences between human and macaque hearts not only enriches our understanding of fundamental cardiac mechanisms, but also opens up new avenues for health and medical research. Such a comparative approach is essential for developing more targeted and effective interventions, thereby helping to improve cardiovascular health in human populations.

Chapter 2: Hormonal Regulation Mechanisms

The regulation of blood pressure and cardiovascular function is a complex process involving interactions between various hormonal systems. Among these mechanisms, the renin-angiotensin-aldosterone system (RAAS) and the action of hormones such as adrenaline play a crucial role. This chapter explores the similarities and differences between these systems, as well as the impact of hormones on cardiovascular regulation.

Renin-angiotensin-aldosterone system (RAAS)

1. How the RAAS works

The renin-angiotensin-aldosterone system is a key hormonal mechanism that regulates blood pressure and fluid balance in the body. It consists of several stages:

- **Renin secretion**: When there is a drop in blood pressure or a fall in blood volume, the juxtaglomerular cells of the kidneys secrete an enzyme called renin. This secretion can also be stimulated by activation of the sympathetic nervous system or by a decrease in sodium in the distal tubule.
- **Formation of angiotensin I**: Renin catalyses the conversion of angiotensinogen, a protein produced by the liver, into angiotensin I, an inactive peptide.

- **Conversion to angiotensin II**: Angiotensin I is then converted to angiotensin II by the angiotensin-converting enzyme (ACE), mainly in the lungs. Angiotensin II is a powerful vasoconstrictor with significant effects on blood pressure.
- **Release of aldosterone**: Angiotensin II stimulates the secretion of aldosterone by the adrenal glands. Aldosterone promotes the reabsorption of sodium and water in the kidneys, thereby increasing blood volume and blood pressure.

2. Similarities between humans and macaques

In humans and macaques, there are notable similarities in the way the RAAS functions. In both species, renin regulation is influenced by factors such as blood pressure, blood volume and sodium concentration. The physiological responses induced by angiotensin II and aldosterone, such as vasoconstriction and sodium reabsorption, are also similar, underlining the importance of the RAAS in blood pressure control.

3. Differences between humans and macaques

However, there are also differences between species. For example, studies have shown that the sensitivity of angiotensin II receptors can vary, which may influence the effectiveness of the vasoconstrictor response and aldosterone secretion. In addition, variations in basal levels of renin and aldosterone have been observed, which may reflect evolutionary adaptations linked to the different lifestyles of macaques and humans.

Role of hormones in regulation

1. Adrenalin and noradrenalin

Adrenaline and noradrenaline, hormones secreted by the adrenal medulla in response to stressful stimuli, play a central role in regulating heart rate and blood pressure.

- **Effects of adrenaline**: Adrenaline increases heart rate, improves myocardial contractility and dilates blood vessels in skeletal muscle, thereby promoting the supply of blood and oxygen to tissues. On the other hand, it causes vasoconstriction in other areas, such as the digestive organs, directing blood flow to the muscles.
- **Effects of noradrenaline**: Noradrenaline, mainly released by the sympathetic nervous system, acts as a powerful vasoconstrictor, increasing peripheral resistance and, consequently, blood pressure. It also plays a role in increasing heart rate and myocardial contractility, but with a slightly different impact to adrenaline.

2. Hormonal interactions

The effects of adrenaline and noradrenaline do not occur in isolation. Their action is often modulated by the RAAS. For example, an increase in angiotensin II can potentiate the vasoconstrictor effect of noradrenaline, creating a more robust cardiovascular response to stress. In addition, RAAS hormones can influence adrenaline release, integrating these systems for effective regulation of blood pressure and heart rate.

3. Adapting to stress

In macaques, activation of the sympathetic system by adrenaline and noradrenaline is particularly important in dynamic social and environmental contexts. Rapid hormonal responses enable efficient adaptation to situations such as competition for food or escape from predators. Differences in hormonal reactivity between macaques and humans may also provide clues as to how evolution has shaped physiological responses to specific environmental pressures.

Conclusion

Hormonal mechanisms, in particular the renin-angiotensin-aldosterone system and the action of hormones such as adrenaline and noradrenaline, play a central role in the regulation of blood pressure and cardiovascular function. Although there are similarities between humans and macaques, major differences in sensitivity and response to hormones underline the importance of evolutionary adaptation. An in-depth understanding of these comparative mechanisms may enrich cardiovascular health research and provide insights into the treatment of cardiovascular disease in humans.

Chapter 3: Nervous Responses

Blood pressure regulation is closely linked to the function of the autonomic nervous system (ANS), which plays an essential role in the rapid adjustment of cardiovascular responses. This chapter focuses on the ANS and the importance of baroreceptor reflexes in humans and macaques.

Autonomic nervous system and its impact on blood pressure

The autonomic nervous system is divided into two branches: the sympathetic nervous system, which prepares the body for stressful situations, and the parasympathetic nervous system, which promotes relaxation and recovery. These two systems work in tandem to regulate blood pressure according to physiological needs.

When a macaque or human faces stress, the sympathetic ANS is activated, leading to an increase in heart rate and vasoconstriction, which raises blood pressure. Conversely, when the threat disappears or the body rests, parasympathetic tone increases, allowing heart rate to decrease and vasodilation to occur, helping to restore blood pressure to normal levels.

Study of baroreceptor reflexes in both species

Baroreceptors are pressure-sensitive receptors located mainly in the carotid artery and aorta. They detect variations in blood pressure and send signals to the central nervous system to adjust the cardiovascular response.

In humans, baroreceptors play a crucial role in maintaining homeostasis. For example, a rapid rise in blood pressure triggers a reflex response that inhibits sympathetic activity and stimulates parasympathetic tone, thereby reducing blood pressure. Studies show that this response is rapid, acting in a matter of seconds to correct abnormalities (Mancia et al., 2007).

In macaques, although the mechanism is similar, research indicates that their ability to regulate blood pressure via baroreceptors can be influenced by factors such as social stress and physical activity. For example, studies have observed that during intense social interactions, macaques can show an altered baroreceptor response, leading to fluctuations in blood pressure (Harrison et al., 2013).

Conclusion

In summary, the autonomic nervous system, via the baroreceptors, plays a fundamental role in the regulation of blood pressure in both humans and macaques. Differences in baroreceptor response can provide valuable information about evolutionary and behavioural adaptations, and their study continues to be a promising area for cardiovascular health research.

Chapter 4: Environmental and Behavioural Factors

Blood pressure is influenced not only by internal physiological mechanisms, but also by a variety of environmental and behavioural factors. This chapter examines the impact of diet and lifestyle on blood pressure, as well as the effects of stress and physical activity.

Influence of diet and lifestyle on blood pressure

1. Food composition

Diet plays a fundamental role in regulating blood pressure, and various studies have established clear links between eating habits and the prevalence of hypertension. Specific components of the diet, such as sodium and potassium, as well as the consumption of fruit and vegetables, have a significant impact on cardiovascular health.

Sodium

Sodium is an essential mineral that plays a key role in maintaining water and electrolyte balance in the body. However, excessive sodium intake is one of the main factors contributing to high blood pressure. Here's how it works:

- **Water retention**: An increase in sodium intake causes the kidneys to retain water. This happens because sodium attracts water, increasing blood volume. Increased blood volume exerts additional pressure on the walls of the blood vessels, raising blood pressure.

- **Recommendations**: Health organisations such as the American Heart Association recommend limiting sodium intake to less than 2,300 mg per day, and ideally to around 1,500 mg per day for people at risk of hypertension. These limits are designed to reduce the risk of cardiovascular disease and stroke.
- **Sources of sodium**: Sodium is often consumed in excess through processed foods, ready meals and salty snacks. Avoiding these sources in favour of fresh foods can contribute to better blood pressure management.

Potassium

Unlike sodium, an adequate intake of potassium can play a protective role against hypertension. The beneficial effects of potassium on blood pressure can be seen in several ways:

- **Balance with sodium**: Potassium helps to balance the effects of sodium in the body. It promotes the excretion of sodium by the kidneys, which can reduce water retention and, consequently, lower blood pressure.

- **Vasodilation**: Potassium also promotes vasodilation, i.e. the relaxation of blood vessels. This reduces vascular resistance and helps to lower blood pressure.
- **Food sources**: Foods rich in potassium include bananas, avocados, sweet potatoes, spinach and beans. Including these foods in your daily diet can help maintain healthy blood pressure.

A diet rich in fruit and vegetables

Diets rich in fruit and vegetables, such as the DASH (Dietary Approaches to Stop Hypertension) diet, are strongly associated with lower blood pressure levels. These foods have a number of health benefits:

- **Rich in nutrients**: Fruit and vegetables are generally rich in vitamins, minerals, antioxidants and fibre, all of which are essential for cardiovascular health. For example, antioxidants can protect blood vessels from oxidative damage, while fibre helps regulate lipid metabolism.
- **DASH diet**: The DASH diet recommends eating several portions of fruit and vegetables a day, as well as whole grains, low-fat dairy products and lean proteins. Studies have shown that people following this diet have a significant reduction in blood pressure.

- **Long-term effects**: A diet rich in fruit and vegetables not only improves blood pressure in the short term, but also contributes to better cardiovascular health in the long term, reducing the risk of heart disease and stroke.

In short, dietary composition has a major impact on blood pressure. Limiting sodium intake while increasing potassium intake and eating a diet rich in fruit and vegetables can promote healthy blood pressure and reduce the risk of hypertension.

Recommended doses :

Sodium

- **Maximum recommended intake** : Limiting sodium intake to less than **2,300 mg a day** is recommended by health bodies such as the American Heart Association.
- **Ideal intake**: For people at risk of hypertension, an ideal intake would be around **1,500 mg a day**.
- **Sources**: A large proportion of sodium intake comes from processed foods and ready meals. Reducing consumption of these products is essential if we are to comply with these recommendations.

Potassium

- **Recommended intake** : The recommended daily intake of potassium for adults is around **2,500 to 3,000 mg per day**. However, some recommendations suggest up to **4,700 mg a day** for optimum effect on blood pressure.
- **Sources**: Foods rich in potassium include :
  - Banana: approximately **422 mg per fruit**
  - Avocado: approximately **975 mg per fruit**
  - Sweet potato: approximately **540 mg per tuber**
  - Spinach (cooked): approx. **540 mg per cup**
  - Beans (cooked): around **600 mg per cup**

A diet rich in fruit and vegetables

- **Recommended portions**: Diets such as DASH recommend eating at least **4 to 5 portions of fruit** and **4 to 5 portions of vegetables** a day.
  - **Portion**: A portion of fruit or vegetable generally corresponds to one cup of raw fruit or vegetables, or half a cup of cooked vegetables.

To sum up, for effective blood pressure management, it is advisable to limit sodium intake to less than 2,300 mg per day, while aiming for a potassium intake of 2,500 to 4,700 mg per day, by including a variety of fruit and vegetables in the daily diet.

2. Lifestyle

An individual's overall lifestyle also influences blood pressure.

- **Alcohol consumption**: Excessive alcohol consumption is linked to an increase in blood pressure. Studies show that moderate consumption (up to one glass a day for women and two for men) can have protective effects, while excessive consumption can lead to hypertension.
- **Body weight**: Overweight and obesity are well-established risk factors for hypertension. Excess weight increases vascular resistance and can also affect hormonal regulation of blood pressure.

Effects of stress and physical activity

1. Stress

Stress, whether physical or emotional, can have harmful effects on blood pressure.

- **Reaction to stress**: In stressful situations, the body activates the sympathetic nervous system, releasing hormones such as adrenaline and noradrenaline. This leads to an increase in heart rate and blood pressure, which can be beneficial in the short term.

However, prolonged exposure to stress can lead to chronic hypertension.

- **Psychological impact**: Psychological stress, such as anxiety or depression, is also associated with an increased risk of hypertension. The underlying mechanisms may involve altered hormonal responses and negative health behaviours, such as unbalanced diet and reduced physical activity.

2. Physical activity

Regular physical activity is a major protective factor against hypertension.

- **Immediate and long-term effects**: Physical exercise causes vasodilation and a temporary increase in blood flow, but over the long term it helps to lower resting blood pressure. Studies show that even moderate activities, such as brisk walking, can significantly reduce blood pressure levels.
- **Mechanisms**: Physical activity improves endothelial function, increases insulin sensitivity and promotes better hormone regulation, all factors that contribute to lower blood pressure.
- **Social and environmental effects**: Engaging in physical activity can also encourage positive social interaction and reduce stress, creating a virtuous circle that promotes cardiovascular health.

Conclusion

Environmental and behavioural factors play a crucial role in regulating blood pressure. A balanced diet, a healthy lifestyle, stress management and regular physical activity are key to maintaining optimal blood pressure. By understanding these influences, it is possible to develop effective prevention and treatment strategies to combat hypertension and improve cardiovascular health.

Chapter 5: Pathological models

Hypertension is a pathological condition that affects both humans and macaques, and represents an essential field of study for understanding cardiovascular disease. By exploring the similarities and differences in the presentation and treatment of hypertension between these two species, we can gain a better understanding of the underlying pathophysiological mechanisms and the effectiveness of therapeutic interventions.

Hypertension in humans and macaques

Hypertension, defined as systolic blood pressure greater than 130 mmHg or diastolic blood pressure greater than 80 mmHg, is one of the leading causes of cardiovascular disease worldwide (WHO, 2021). In humans, several risk factors contribute to this condition, including:

- **Diet**: High consumption of sodium and saturated fats.
- **Lifestyle**: Lack of physical exercise and excess weight.
- **Genetic factors**: Family history of hypertension.
- **Stress**: Psychosocial factors contributing to high blood pressure.

In macaques, research has shown that environmental and behavioural factors, such as socialisation, diet and level of physical activity, also influence the development of hypertension. Experimental studies have shown that exposure to sodium-rich diets and levels of psychosocial stress can induce hypertension in these primates (Kraus et al., 2008).

Macaques can develop cardiovascular diseases similar to those of humans, including vascular lesions and cardiac dysfunction, making them a useful model for research.

Available data on the prevalence of hypertension in macaques shows an incidence similar to that observed in humans, with estimates of up to 50% in certain groups of ageing macaques (Kraus et al., 2008). These models make it possible not only to study the progression of hypertension, but also to assess the pathophysiological mechanisms involved.

Therapeutic interventions and their effectiveness

Interventions to treat hypertension in humans include lifestyle changes, antihypertensive drugs and behavioural approaches. Lifestyle changes include:

- **Diet**: reducing sodium intake, increasing potassium consumption and adopting diets such as DASH (Dietary Approaches to Stop Hypertension).
- **Physical exercise**: Regular activity recommended to lower blood pressure.
- **Stress management**: relaxation techniques, meditation and behavioural therapy.

As far as drugs are concerned, several classes of antihypertensive are commonly used:

- **Diuretics**: Help eliminate excess sodium and water.
- **Angiotensin-converting enzyme inhibitors (ACEI)**: Reduce the production of angiotensin II, a powerful vasoconstrictor.
- **Beta-blockers**: Reduce heart rate and the force of cardiac contraction.

Interventions in macaques follow similar principles. Studies have shown that dietary modifications, such as reducing sodium intake and increasing potassium intake, lead to significant improvements in blood pressure (Hoffman et al., 2004). In addition, the use of antihypertensive drugs has been successfully tested in macaques, showing promising results in terms of reducing blood pressure and improving cardiovascular function (Reid et al., 2015).

Conclusion

The study of hypertension in humans and macaques, and the evaluation of therapeutic interventions, offer essential insights into understanding this complex condition and improving treatment strategies. By using macaques as models, researchers can explore new therapies and adapt clinical approaches to better meet the needs of hypertensive patients.

Chapter 6: Evolution and Adaptation

The evolution of blood pressure regulation mechanisms and species-specific adaptations to their habitats and lifestyles provide a fascinating insight into how organisms have evolved to maintain cardiovascular homeostasis in the face of varied environmental challenges. This chapter explores these two essential dimensions.

Evolution of blood pressure regulation mechanisms

Over the course of evolution, blood pressure regulation mechanisms have undergone modifications to adapt to specific environments and the physiological needs of different species. The earliest forms of life had simple circulatory systems, but as organisms became more complex, the need for effective blood pressure regulation became paramount.

Mammals, including humans and macaques, have a sophisticated hormonal regulatory system, notably the renin-angiotensin-aldosterone system (RAAS), which plays a key role in managing blood pressure. Evolution has made it possible to optimise this system in response to a variety of conditions, such as fluctuations in sodium intake and water requirements. For example, in species living in arid environments, adaptations such as increased water retention and more efficient sodium regulation have been observed (Guyton, 1991).

In addition, the evolution of the autonomic nervous systems, in particular the modulation of sympathetic and parasympathetic tone, has also been essential for the adaptation of species to stressful situations. These systems enable rapid and appropriate responses to external stimuli, ensuring adequate blood pressure in a variety of contexts, such as flight from a predator or complex social behaviours (Reid et al., 2015).

Specific adaptations to habitats and lifestyles

Adaptations to blood pressure regulation mechanisms are often closely linked to species' habitats and lifestyles. For example, aquatic species, such as certain fish, have developed unique strategies for maintaining stable blood pressure in an environment where pressure changes frequently. Their circulatory systems are often less complex, but are extremely efficient at coping with the challenges of life in water.

Primates, on the other hand, show more complex adaptations. Macaques, for example, which evolve in a variety of habitats ranging from tropical forests to mountainous regions, have regulatory mechanisms that enable them to adapt to rapid changes in their environment. Their higher heart rate and ability to manage stress effectively contribute to their survival in conditions where competition for resources is high (Harrison et al., 2013).

What's more, diet plays a crucial role in these adaptations. Species that eat a diet rich in potassium, such as fruit and vegetables, tend to have more stable blood pressure and better regulation of body fluids. This contrasts with species that consume a diet rich in sodium, where more intense hormonal regulation is required to compensate for water retention.

In short, the evolution of blood pressure regulation mechanisms and the specific adaptations to habitats and lifestyles illustrate the complexity of the physiological responses of species to their environments. These adaptations, whether hormonal, behavioural or dietary, are essential to ensure the survival and well-being of organisms in a constantly changing world.

Conclusion

The comparative study of blood pressure regulation mechanisms in humans and macaques has brought to light significant findings with far-reaching implications for future research and cardiovascular health.

Summary of discoveries

Throughout this work, we have explored the anatomical, physiological and hormonal differences between these two species, highlighting how evolutionary adaptations and environmental influences modulate the cardiovascular response. Regulatory mechanisms, including the renin-angiotensin-aldosterone system and the interaction between the sympathetic and parasympathetic nervous systems, were shown to be crucial in controlling blood pressure. In addition, behavioural factors such as diet and stress have been shown to play an essential role in the development of hypertension.

This research also highlights the importance of macaques as models for studying human cardiovascular pathologies, offering unique insights into pathophysiological mechanisms and therapeutic interventions.

Implications for future research

The results of this study pave the way for new investigations into the complex interactions between genetic, environmental and behavioural factors in the regulation of blood pressure. Future research could focus on :

- **Longitudinal studies**: Analyse how environmental changes influence blood pressure over time.
- **Clinical and experimental trials**: Testing new pharmacological or behavioural interventions in animal models prior to their clinical application.
- **Translational research**: Assessing how discoveries made in macaques can be applied to improve the treatment of hypertension in humans.

Perspectives on cardiovascular health and translational research

Worldwide, hypertension remains one of the main contributors to cardiovascular disease, with significant consequences for public health. Understanding the mechanisms that regulate blood pressure, particularly using animal models, may lead to innovative approaches to the prevention and treatment of hypertension.

Translational research, which links laboratory discoveries to clinical applications, is essential for developing effective public health strategies. By integrating knowledge acquired from animal models, such as macaques, researchers can design interventions that are better adapted to the specific needs of human populations, taking into account genetic and environmental variations.

In conclusion, this study highlights the complexity of the mechanisms regulating blood pressure and the importance of interdisciplinary research in improving our understanding and ability to treat hypertension. The discoveries made pave the way for a promising future in cardiovascular research, with an emphasis on collaboration between scientific disciplines to translate this knowledge into concrete benefits for human health.

Glossary :

1. **Cardiac anatomy**: Study of the structure and shape of the heart and its components.
2. **Atrium**: Upper chamber of the heart which receives blood from the veins, with a right and a left atrium.
3. **Ventricle**: Lower chamber of the heart which pumps blood to the lungs (right ventricle) or to the rest of the body (left ventricle).
4. **Cardiac conduction system**: set of specialised structures (sinus node, atrioventricular node) which control the rhythm and synchronisation of cardiac contractions.
5. **Myocardial mass**: Weight of the heart muscle, varying according to physical activity and metabolic needs.
6. **Myocardial perfusion**: supply of blood to the heart muscle, essential for its optimal functioning.
7. **Heart rate (HR)**: Number of heartbeats per minute, a key indicator of cardiovascular health.
8. **Cardiac output**: Volume of blood pumped by the heart per minute, determined by heart rate and ejection volume.
9. **Autonomic nervous system (ANS)**: Part of the nervous system controlling involuntary functions, including heart rate, digestion and breathing.

10.**Sympathetic nervous system**: Branch of the ANS which prepares the body for stressful situations ("fight or flight"), increasing heart rate and myocardial contractility.

11.**Parasympathetic nervous system**: Branch of the ANS that promotes recovery and relaxation and lowers the heart rate after a period of stress.

12.**Myocardial contractility**: Capacity of the heart muscle to contract and pump blood.

13.**Parasympathetic tone**: Level of activity of the parasympathetic nervous system, influencing heart rate and recovery.

14.**Homeostasis**: A state of dynamic equilibrium of physiological processes in the body, essential for survival.

15.**Recovery**: Process by which the body returns to a state of rest after a period of physical activity or stress.

16.□ **Translational research**: Scientific process linking basic research to clinical practice, aimed at applying scientific discoveries to improve healthcare.

17.□ **Basic research**: scientific studies aimed at understanding biological mechanisms, with no immediate clinical application in view.

18.□ **Applied research**: Studies that seek to solve concrete problems or develop technologies and treatments based on theoretical knowledge.

19. **Clinical trials**: Research studies carried out on human participants to assess the efficacy and safety of new treatments, drugs or medical devices.

20. **Translational research continuum**: model illustrating the different stages of research, from discovery in the laboratory through preclinical development to clinical application.

21. **Medical innovation**: Introduction of new ideas, methods or devices that improve the diagnosis, treatment or prevention of disease.

22. **Biomarker**: A measurable biological indicator that helps to detect a disease, assess disease progression or predict response to treatment.

23. **Personalised medicine**: A therapeutic approach that adapts treatments to the individual characteristics of patients, often based on genetic or biological data.

24. **R&D (Research and Development)**: The process of research and experimentation to develop new products, technologies or methods.

25. **Public-private partnerships**: Collaborations between the public sector (governments, research institutions) and the private sector (companies, industries) to foster innovation and the development of new healthcare solutions.

References :

1. Appel, L. J., et al. (1997). "A clinical trial of the effects of dietary patterns on blood pressure." *New England Journal of Medicine*, 336(16), 1117-1124.
2. Buchanan, J. W., & Becker, J. A. (2010). "The renin-angiotensin system: Physiology and pharmacology." *Cardiovascular Pharmacology*, 54(3), 235-245.
3. Cornelissen, V. A., & Smart, N. A. (2013). "Exercise training for blood pressure: a systematic review and meta-analysis." *Journal of the American Heart Association*, 2(1), e004473.
4. Dunbar, R. I. M. (2009). *The social brain: mind, language, and society in evolutionary perspective.* Annual Review of Anthropology, 38, 211-228.
5. Geleijnse, J. M., et al. (2005). "Dietary approaches to prevent hypertension: the role of potassium." *The American Journal of Clinical Nutrition*, 81(5), 1058-1067.
6. Harrison, D. G., & Coffman, T. M. (2003). "The renin-angiotensin system in hypertension: a new perspective." *The Journal of Clinical Investigation*, 111(7), 1017-1025.
7. Harrison, G. A., et al. (2013). "Social stress and blood pressure regulation in primates." *American Journal of Primatology*, 75(12), 1196-1203.

8. Hoffman, J. R., et al. (2004). "Physiological responses of the macaque to exercise and its implications for understanding human cardiovascular function." *Journal of Physiology*, 559(2), 561-570.

9. Koch, W. J., & Lefkowitz, R. J. (2005). "Adrenergic receptor signaling in the heart: Regulation by the renin-angiotensin system." *Circulation Research*, 96(1), 82-93.

10. Kraus, W. E., et al. 2008. "Effects of a high-sodium diet on blood pressure and cardiovascular health in nonhuman primates." *Journal of Hypertension*, 26(9), 1815-1822.

11. Lindsey, L. A., et al. (2014). "Comparative anatomy of the coronary arteries in primates." *American Journal of Primatology*, 76(6), 566-578.

12. López-Sepúlveda, R., et al. (2016). "The Role of the Renin-Angiotensin System in the Regulation of Cardiovascular Function." *Frontiers in Physiology*, 7, 292.

13. Mancia, G., et al. (2007). "Blood pressure regulation: The role of baroreceptors." *Hypertension*, 49(3), 569-576.

14. Reid, J. W., et al. (2015). "The influence of the autonomic nervous system on heart rate variability in primates." *Physiological Reports*, 3(7), e12473.

15. Reid, J. L., et al. (2015). "The Role of Stress in Blood Pressure Regulation: Lessons from Animal Models." *Hypertension Research*, 38(3), 185-192.

16. World Health Organization (WHO). (2021). "Hypertension." Retrieved from WHO website.

17. Whelton, P. K., et al. (2018). "Guideline for the Prevention, Detection, Evaluation, and Management of High Blood Pressure in Adults." *Journal of the American College of Cardiology*, 71(19), e127-e248.

Printed by Books on Demand GmbH, Norderstedt / Germany